Kindergarten Polar Bear Unit Study

Author:
Sarah Bean
Raisinghumanbeans.com

How to use this Study

In each Science study, there are 2-3 math pages; vocabulary, sight words, and a myriad of other activities and pages.

These Units were designed to take one week, in monthly conjunction with our State and President Studies; but you can extend the activities if you wish. I will soon have additional math lessons for purchase, and we are working on developing a Big Book of Unit Study, for K-6.

If you do follow the week approach, I recommend Math on MWF, Vocab and Sight Words Daily, and the other Activities Daily.

This is an open and go book, but planning for me always works better.

Good luck on your Homeschooling Journey!!

Note:
There are some pages that mention to cut out. Please copy these pages if you wish to cut them out, so you still have the pages on the back.

Table of Contents

Page Number

- 3-4   Sight Words
- 5-6   Vocabulary
- 7-8   Plant Facts
- 9   Lines
- 10-12   Tracing
- 13-15   Parts of a Polar Bear
- 16-17   Lifecycle
- 18   Matching
- 19-21   Math
- 22   5 Senses Hunt
- 23   Bingo
- 24-26   Tic Tac Toe
- 27-28   Recipe

SIGHT WORDS

Bear	White
Cold	Meat

SIGHT WORDS

WEEK 1

For the first week, show these to your child every day. Make sure they are looking at the card, and go through all of them 3 times. Don't ask them to tell you what it is, just tell them what it is.

WEEK 2

1st round- Ask them what the words are. If they don't know, tell them.

2nd round- Put them on the ground in a circle. Ask them to jump to "this word." See if they are right!

3rd round- If they are still having trouble, explain to them why the word is the word. Have them remember based on sounding out the first few letters. Play the matching game a few times!

VOCABULARY

Carnivore:
This means that Polar bears only eat meat.

Extreme:
Very great, the greatest. Extreme cold; the greatest cold.

VOCABULARY

Practice these daily.

You can learn none of these; one of these; or all of these.

If you learn 1 or 2, this Unit could take 1 week.

If you learn 3-5, this Unit could take up to a month.

Remember: Go at your child's pace.

POLAR BEAR FACTS

1. Polar Bears are Carnivores.

2. They feed mostly on seal, but will also eat fish, small mammals and birds.

3. Polar bears are not white, they are transparent. Their fur is meant as camoflauge.

POLAR BEAR FACTS

4. They usually only have 2 cubs.

5. They live in extremely cold temperatures, -32* below.

6. Polar Bears are Endangered, which means there are very few left.

POLAR BEAR LINES

Stay on the line, try to get the Bear to the other side!

Trace the Polar Bear, give him ears and eyes and whatever else you like!
Take a picture and send them to me to be featured :)

Trace the Snowflake, and give it some kind of design! Lines, circles, whatever you like! Take a picture and send them to me to be featured :)

Trace the word Bear!

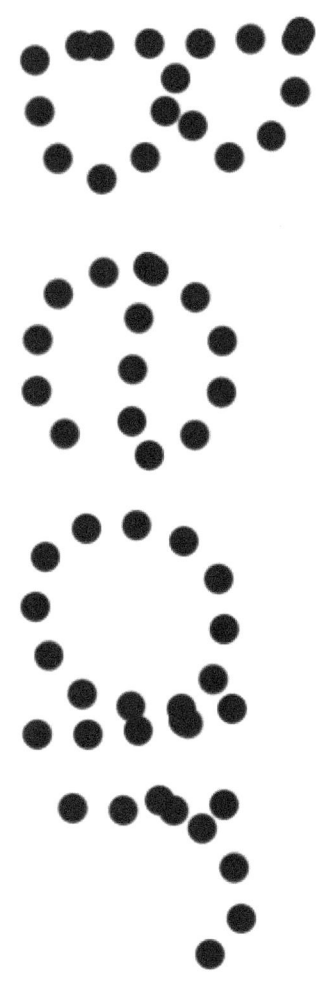

PARTS OF A POLAR BEAR

PARTS OF A POLAR BEAR

Round Ears

Muscular hump

Claws

Fur

Cut and paste them in the boxes above.

PARTS OF A POLAR BEAR

1. The muscular hump helps them in catching prey.
2. Large claws help them eat their prey.
3. Fur helps them camouflage in the snow.
4. Polar bears hearing isn't the best, but their sense of smell is amazing.

POLAR BEAR LIFE CYCLE

4

3

1

2

POLAR BEAR LIFE CYCLE

1. The baby is born
2. Mama teaches baby to hunt
3. Baby grows, digs a hole for the winter.
4. Mama has babies while sleeping.

Cut these pictures out and place them in order in your own life cycle!

POLAR MATCHING

   White

   Bear

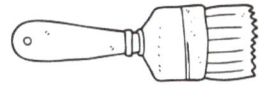   Cold

   Meat

POLAR MATH

- Place an object on each part of the snowflake. Try to keep them in the lines!

POLAR MATH

Draw a line from the answer to the problem.

POLAR MATH

Count how many there are in each group, write the number next to the group.

5 SENSES

Go on a Scavenger Hunt. This is best do to in the snow. Find something that looks like a Polarbear, or where one would live, tastes like what a polar bear eats, sounds like a polar bear, feels like a polar bear, and smells like a polar bear. X out when you are done. Be creative!!

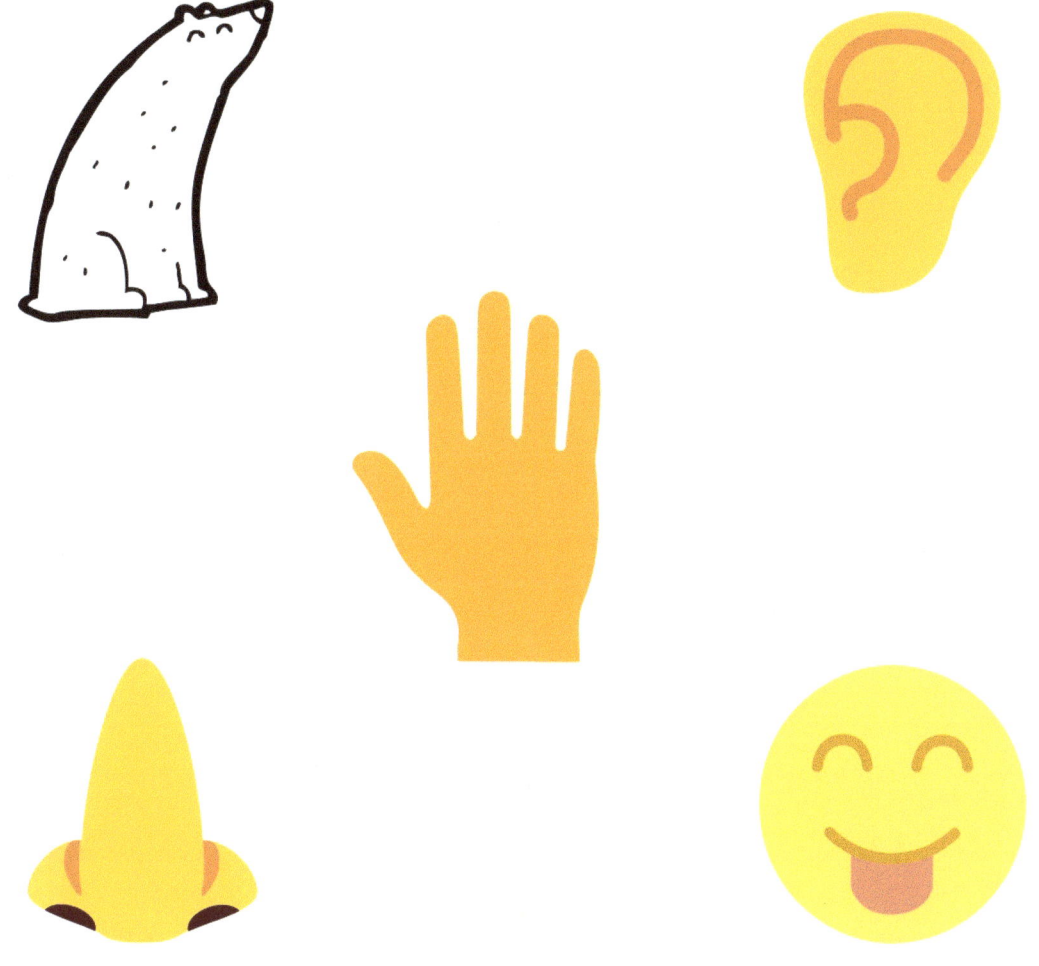

POLAR BEAR BINGO

Roll a die. Count the dots, and put a marker on the corresponding picture. Play til someone wins!
You may need to roll the die twice, if so it is a good math lesson! Addition.

TIC TAC TOE

These are your x's and o's. Cut them out, and you can have up to 3 players.

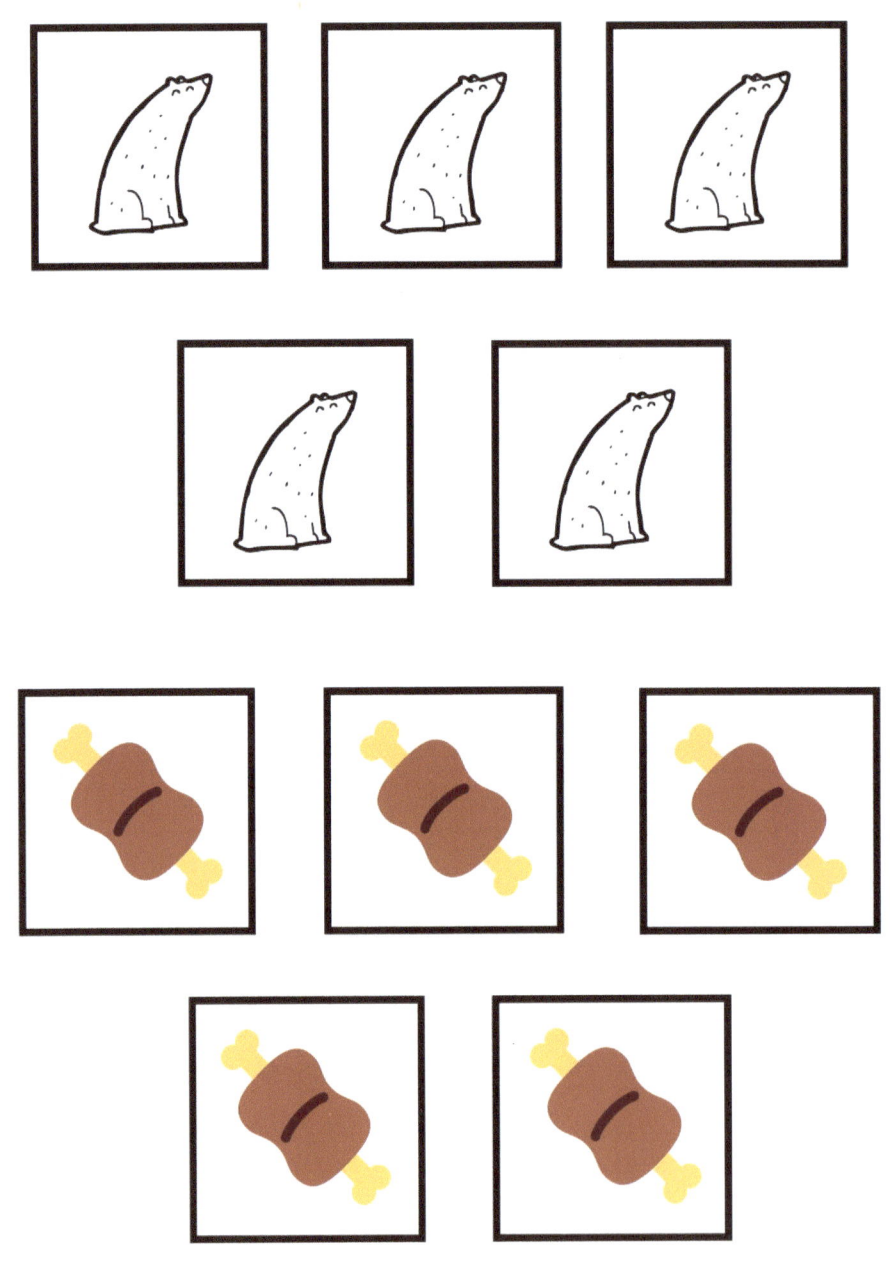

ICE COLD RECIPE

Homemade Ice Cream

Ingredients:
2 cups Heavy Cream
1, 14oz can Sweetened Condensed Milk

Optional Mixins:
Vanilla extract
Peppermint extract
Candy bits
Brownie or cookie dough
Peanutbutter or Chocolate Syrup
Fruit pieces

ICE COLD RECIPE

Homemade Ice Cream

Directions:

Whip COLD cream until peaks form.

Add condensed milk, and whip until peaks form again!

Then, add any mixins.

Freeze for 4-6 hours or overnight!

Follow us!

Raisinghumanbeans.com

Facebook.com/raisingbeans

Instagram.com/raisinghumanbeans

Pinterest.com/raisinghumanbeans

© Raising Human Beans 2018

www.ingramcontent.com/pod-product-compliance
Lightning Source LLC
Chambersburg PA
CBHW042323250526
R18347300001B/R183473PG45473CBX00017B/9